MAORI WEAPONS
in Pre-European New Zealand

Jeff Evans

REED

Established in 1907, Reed Publishing (NZ) Ltd
is New Zealand's largest book publisher, with over 300 titles in print.

For details on all these books visit our website:
www.reed.co.nz

Published by Reed Books, a division of Reed Publishing (NZ) Ltd, 39 Rawene Rd,
Birkenhead, Auckland. Associated companies, branches and representatives
throughout the world.

ISBN 0 7900 0826 2

Edited by Peter Dowling
Designed by Craig Violich
Unless credited, photos are by Jeff Evans

First published 2002

Printed in New Zealand

CONTENTS

FOREWORD

Interest in Maori weaponry is growing at present, especially through use of taiaha and other traditional weapons in kapa haka and cultural group performances, as well as through organised marae-based traditional weapons classes.

The representations of Maori weaponry are as diverse as the fighting styles used in the various tribal areas throughout Aotearoa. Use of weapons as an assertion of tribal power contrasts, for example, with their place in portraits of Maori as 'noble savages'.

It seems that Maori weaponry embodies an authority and power, a mana that is attractive to young Maori eager to explore and reclaim their identities. Perhaps also, the taiaha represents the traditional knowledge and connection with the natural world of ancient times.

That knowledge was part of a comprehensive historical recollection that could relate weapons to events or people, thereby providing a spiritual focus for the weapons. Naming a weapon in connection to a friend or foe established a special relationship that would enhance its mauri or life principle, a vital link for items so important to daily life and death.

Despite the positive aspects of the re-emergence of traditional Maori weaponry, weapons in general are nonetheless a symbol of violence that highlights some of the tensions between cultural pride and the negative family violence statistics in Maori homes.

All of these things are important considerations to people like myself who make, teach and use Maori weapons today. Any activity which brings to life the unique Maori identity and helps us forward out of the mire of colonisation must be supported. I welcome the respectful exploration of traditional knowledge in the creation of weapons and other art forms that I am familiar with, such as working with wood, bone, shell, fibres and dyes.

Maori Weapons is one such exploration. By way of this introductory volume, Jeff Evans presents much valuable information about Maori weapons and weaponry. I commend this compilation as a good starting point for those interested in the broad spectrum of Maori weaponry. It is more than a purely academic work or a regurgitation of previously published material. His work attempts to span the gap between the historical and contemporary worlds in terms of Maori weapons.

It remains for those involved in making, teaching and using these weapons to design the appropriate context for these activities, to ensure that this information is used as a leverage for cultural progress and not merely as a means of cultural destruction.

Bernard Makoare
Ngati Whatua
July 2002

ACKNOWLEDGEMENTS

I would like to acknowledge and thank a number of individuals who have assisted in the completion of this book.

Firstly my sincere thanks to Roger Neich, Kelvin Day and Hirini Melbourne, who have each read over the manuscript at various stages and made important suggestions to improve and strengthen the content. The finished book is richer for their input.

Once again I am indebted to Peter Janssen and Peter Dowling of Reed Publishing who have pushed and prodded me through to the completion of yet another book. Thank you for your encouragement and patience.

To Pita Sharples and Paora Sharples, thank you both for your willingness to share your knowledge with me when I have sought your expert opinion on a wide range of points, as well as allowing me to sit in on numerous mau rakau classes at Hoani Waititi Marae.

Finally a special thank you to Bernard Makoare for the time and effort spent writing the foreword and reading over the manuscript. Attendance at your taiaha-making wananga was both invaluable to my research and a pleasure to experience. I thank you sincerely.

And to my beautiful wife, thank you once again for your ongoing support and encouragement.

Jeff Evans
July 2002

AUTHOR'S NOTE

Information about the weapons described in this book has been gathered from written accounts from all parts of Aotearoa, many of them recorded in the early years of Maori–Pakeha contact. There are inevitably gaps in the information that is available — some of the weapons discussed, such as the taiaha, are fairly well documented, while others, including the hoeroa and patuki, are more scantily addressed.

It should also be noted that variations to the techniques and styles described in the book are known to have existed, and there were frequently individual and tribal adaptations to all facets of weaponry. Such variations often resulted from personal preferences, although the limited availability of favoured materials, or perhaps the environment in which the weapon was to be used, were also contributing factors. Unfortunately, notes relating to such modifications, if they ever existed, are no longer to be found.

While every effort has been made to acknowledge tribal origins for the weapons illustrating this book, surprisingly little information of this type has been collected or recorded. Where it is available the information and the museum reference number are given.

INTRODUCTION

This book provides an introduction to the weapons used by the Maori warrior, or toa, in the days before Europeans arrived in New Zealand. These are the weapons that were used in the battles and skirmishes that frequently raged in the centuries before the musket was introduced in the early 1800s, changing the nature of warfare for ever.

Weapons and warfare occupied a prominent place in the pre-European landscape of Aotearoa. Over time the Maori mastered all aspects of warfare, including hand-to-hand fighting, ambushes, the siege and defence tactics of pa warfare, and massed attacks between well-armed forces. As in many other parts of the globe, however, it was hand-to-hand fighting that helped settle most personal and inter-tribal feuds. Toa strove for perfection in the skills required to survive hand-to-hand combat, and often sought out skilled adversaries to test themselves against.

Not only did hand-to-hand fighting offer the toa a chance to demonstrate his bravery, skill and strength, it also offered the opportunity for widespread fame. The Maori scholar Te Rangi Hiroa (Sir Peter Buck) wrote of the toa in *The Coming of the Maori*:

The fame of the warrior (toa taua) was held transient as compared with that of the provider of food, yet the fact remains that noted warriors received more publicity in song and story than their contemporaries in the peaceful arts. The fame of noted warriors spread far beyond their tribal bounds and intense rivalry and jealousy existed among the champions of the different tribes. Hence the preliminary staging of single combat between opposing armies. Warriors sought descriptions of their rivals in order that they might identify them on a crowded battlefield.

To attain the prowess needed for the battlefield, boys, and in a few notable cases girls, were taught the skills of warfare from an early age. Each tribe or subtribe ran schools where the skills of war were taught. Beginning with light sticks, students were taught attacking and defensive moves, foot drills and other related skills.

When preparing for battle, the Maori warrior had a choice of three basic types of weapon: short patu and mere, often mistakenly referred to as clubs; longer staff-like weapons, such as the taiaha, which were designed to be held with both hands; and spears, both thrown and hand-held. Among the patu and mere were the patu paraoa, patu onewa and mere pounamu, while two-handed weapons included taiaha, tewhatewha and pouwhenua. Among the spears were the tarerarera and timata, which were designed to be thrown; the long, hand-held spear called the huata; and a number of shorter spears used in hand-to-hand fighting. Other weapons, such as the oka, were less widely available. Stones were also used, and could reportedly be thrown very accurately over great distances, but their use was usually restricted to the defence of pa.

Patu were used as a thrusting weapon, rather than as a club, with the leading edge being the primary striking point, although it was common to finish an enemy off by striking them with the butt of the handle. These short weapons were made from whalebone, wood or stone, including the valued pounamu (nephrite), commonly called greenstone. The only weapon known to be used as a club (that is, with a clubbing motion) was the patuki, a comparatively rare weapon that survives today in a limited number of collections around the world.

The longer, staff-like weapons, such as the taiaha and tewhatewha, were generally held with both hands, and were used to strike at the enemy and to parry attacks. These weapons all featured a pointed end for jabbing, and a flat blade, the edge of which was used for striking.

There were several types of spear. Throwing spears, which are relatively rare in collections of Maori weapons, are known to have been deadly when thrown by skilled warriors, and were often used to good effect. The longer spear known as the huata was used primarily to defend, or on rare occasions attack, fortified pa. The third type of spear was used in hand-to-hand combat, and included the tao and koikoi, both of which were employed to parry an opponent's attack and counter with a strike of its own in much the same way that the English used the halberd.

This impressive wood engraving entitled '*Group of New Zealand chiefs now on a visit to England*' highlights the diversity of weapons carried by Maori. Clearly visible are taiaha, tewhatewha, long spears (probably a tokotoko on the ground and a tao at the centre, rear), and various patu and mere.
(*Penny Illustrated Weekly News* 1863; Alexander Turnbull Library, Wellington, A-018-015)

It is fair to say that uniformity was not a regular feature of Maori weaponry, although the basic profile of weapons was always preserved. The length, weight and size of any weapon was determined in part by the warrior's preference, with weapons customised to each individual, as well as being influenced by the dimensions of the available material; this was particularly the case when working with paraoa (whalebone) and pounamu (greenstone).

Not only was there a lack of uniformity between individual weapons, but any single war party, or taua, usually carried a great variety of weapons. During a tense stand-off in the Bay of Islands, the Frenchman Marion du Fresne witnessed a party of warriors advance towards his position, armed with long spears, 'darts' (probably tarerarera, throwing spears that were on average about 3 m long), 'truncheons' (probably taiaha), and various 'war clubs' (patu), which du Fresne was later told were 'used for splitting skulls'.

Du Fresne also had the opportunity to visit a pa during his stay in the Bay of Islands, where he and a number of his men were given a tour of the village, including a look inside the hut where the arms were stored. They found what du Fresne described as a 'surprisingly large' quantity of small wooden spears, clubs, taiaha, 'tomahawks' (probably patu) of stone and whalebone, throwing darts, and various battle-axes (probably tewhatewha) in hardwood. It is likely that armouries such as this were to be found in most pa in the late 1700s and early 1800s.

On the whole, the limitation of the Maori armoury to weapons made of stone, wood and whalebone had for centuries managed to keep large-scale scenes of bloodshed in check. But the introduction of European firearms resulted in a swift and uncompromising change to this. In much the same way that the introduction of firearms in Japan had led to the demise of the famed samurai warrior, the effectiveness of the Maori toa trained in the arts of traditional weaponry was negated virtually overnight by the arrival of the musket. While traditional weaponry did survive to some extent, the availability of weapons that could kill the most skilled of warriors from a safe distance, even in the hands of a relative novice, changed the face of warfare in Aotearoa for ever.

TWO–HANDED WEAPONS

The two-handed weapons were the taiaha, pouwhenua and tewhatewha. Each of these weapons averaged some 1.5–1.8 m in length, and in most cases they were constructed out of hardwoods such as manuka, maire, akeake or rautangi. These woods were particularly prized for their qualities of strength and relative lightness. Whalebone was occasionally used for such weapons, and a small number of such examples survive today.

The taiaha, pouwhenua and tewhatewha all combined a blade for striking and a proximal point for stabbing, and in the hands of an expert they were deadly. Of the three, the taiaha is perhaps the most common in collections today. It had a long, narrow blade for striking at one end, and a protruding arero, or tongue, for jabbing at the other. The arero, which was often elaborately carved, extended from the upoko, or head, which was also richly carved. Many taiaha in museum and private collections are further decorated with a collar of red feathers just below the upoko, and on the more elaborate, tufts of white or grey dog's hair, or awe, protrude from under the red feathers.

The pouwhenua was an equally effective weapon, with which a skilled warrior could easily dispatch an enemy with a single blow. It often had a wider and heavier blade than the taiaha, and it did not have the elaborate carving seen on the upoko and arero of the taiaha. The sharp proximal point was often smoke- or fire-hardened, and was perfect for finishing off a fallen opponent.

The tewhatewha was a weapon with a distinctive profile, and was often described as axe-like by the first Europeans to visit Aotearoa. Its most obvious feature was the flat, quarter-round section protruding from the top of the otherwise straight blade. This added extra weight behind the blows, which were struck with the straight front edge. Most tewhatewha were decorated with tufts of hawk feathers attached to the quarter-round section of the blade; these were used to distract the enemy as they were flicked across his face, allowing a swift blow to be struck.

All these weapons were usually held along the shaft and used in a series of quick thrusting and parrying movements. A favourite strategy was for the warrior to feint a jab at an enemy with the point of the weapon before reversing it to strike with the blade at the head or shoulders. Sufficient momentum and power could be generated with such a swift turn to kill an enemy outright. When approaching combat, these weapons were usually held vertically or diagonally across the body, with the striking blade above.

POUWHENUA

The graceful pouwhenua was the simplest in form of the double-handed weapons. Although superficially similar in many respects to the taiaha, it lacked the taiaha's carved upoko and arero (head and tongue), instead having a plain, sharp, fire- or smoke-hardened point. Like the taiaha, the striking edge along the pouwhenua's rau, or blade, was comparatively blunt, and best used to smash down on a victim, rather than cutting across in the way a sword might be used. The rau of the pouwhenua was often flared a little more than that of the taiaha.

The pouwhenua narrowed gradually from the wide blade through the shaft

This pair of pouwhenua from the collection at Te Papa demonstrates the flaring of the rau or blade, which is much more pronounced than that on the taiaha (see image on page 20 for a comparison). Note the rare carving halfway up the shaft of the pouwhenua on the right. (Collection of the Museum of New Zealand Te Papa Tongarewa, B.18520, B.18521)

Detail of a typical carving design found on the shaft of pouwhenua.
(Auckland Museum, AM 3480)

to a finely carved band, or whakawhiti, about 40 cm up from the point. From the whakawhiti down the cross-section was rounded, narrowing to the plain point. While the occasional pouwhenua was carved all over, it seems certain that such examples were employed exclusively as ceremonial weapons, or carved for trade with Europeans.

The initial construction sequence was the same for both the pouwhenua and the taiaha. Using a wood such as manuka, akeake or rautangi, the first step was to reduce the diameter of the raw timber through a process of adzing. Starting with heavier toki, or adzes, the wood was rough-hewn to the general size and shape required, then further reduced using a series of finer and lighter toki until the final shape was reached. Stone scrapers were then employed, followed by a rubbing with sandstone and pumice. This process left a slightly rippled effect down the shaft, which is most noticeable if you look down the length of the weapon on an angle to the light. To finish off the construction, the weapon was sometimes rubbed on the trunk of a fern, which resulted in a finely polished surface.

All wooden weapons such as these were oiled occasionally to prevent cracking and to sustain the smooth surface. Human fat was sometimes used for this purpose, as well as the more common shark oil. Like other weapons, pouwhenua were often placed in the rafters of the whare, or houses, where the smoke continued to harden them.

The pouwhenua was wielded in two hands, and employed in much the same manner as the taiaha. As noted above, it had a sharp point for jabbing, while the blade was used to chop down on the opponent's head, shoulders, body or legs. The highly polished finish of the pouwhenua ensured that when used in combat it slid swiftly through the warrior's half-closed hand as he delivered a thrust with the point.

According to evidence collected by the ethnographer Elsdon Best in the late nineteenth and early twentieth centuries, some East Coast Maori called a pouwhenua with a carved band around its shaft a kaukau. Best also identified a weapon called a tarowai, which he thought was the pouwhenua under another name, although there is no other evidence to back this claim.

TEWHATEWHA

The tewhatewha was commonly referred to by Maori as the 'rakau rangatira', or chiefly weapon. This was because it was often seen in the hands of chiefs, either signalling warriors during battle, on the marae, or marking time for paddlers in waka taua. It was particularly effective for this because it stood out physically from all other weapons, and the feathers that usually decorated it added to its visibility.

It is easy to see why the tewhatewha was often termed a battle-axe by early visitors to Aotearoa. While the shaft closes down to a mata, or point, similar to that of the pouwhenua, the main visual feature of the tewhatewha is

This selection of tewhatewha highlights the slight variations typical of the weapon. Of particular note is the curve to the shaft and the concave angle along the top of the quarter-round section. This is also a very good illustration of the positioning and grouping of puhipuhi, or feathers. (Collection of the Museum of New Zealand Te Papa Tongarewa, B.18541)

Taken c. 1930 at Parihaka, this photograph shows a kaumatua holding a short tewhatewha while speaking on the marae. Short tewhatewha like the one depicted here are relatively rare and were possibly designed for ceremonial uses rather than hand-to-hand combat.
(Alexander Turnbull Library, Wellington, G-17415-1/2)

the broad, quarter-round head at the striking end, called the rapa. Despite the tewhatewha's similarity to a European axe, it was the edge of the shaft in front of the rapa, rather than the rounded edge, that was used when striking an enemy. The rapa was there primarily for the extra weight it provided when striking.

The long handle of the tewhatewha was usually oval in cross-section, and often carried some surface carving on a slightly raised boss about 40 cm up the shaft from the mata. The carving often represented two human faces looking outwards, and was possibly added to prevent the hand slipping too far along the shaft during use. The quarter-round rapa tapered to a thin edge, and it was usually slightly concave along the top. Tewhatewha typically averaged 1.1 m in length.

According to Herries Beattie, the favoured woods for making tewhatewha in the South Island were rata, kahikatoa and manuka, while Elsdon Best states that the root of the maire tree was used in the Urewera area. The root of the maire was particularly suited for the construction of tewhatewha, being close-grained and able to withstand a lot of use without splitting. There are also a small number of tewhatewha fashioned from whalebone in collections.

This pair of tewhatewha illustrates two distinctive carving styles used for this weapon. Characteristically, both carvings would feature paua shell discs for the eyes.

Many tewhatewha had a hole drilled through the lower part of the rapa to accommodate small bunches of hawk or kereru feathers, called puhipuhi or taupuhi. The feathers were prepared by being split down the middle, and as much of the quill (tuaka) as possible removed, leaving a thin strip that formed into a spiral. Experienced warriors would often flick the suspended feathers at their opponent's eyes, and while they were distracted poke them with the mata before reversing arms and striking their head with the straight front surface of the blade. If successful the warrior would complete the assault by finishing off his prone adversary with the mata.

While the tewhatewha was not as common as the taiaha or the numerous variations of the patu and mere, its strong aesthetic qualities ensured that it is well represented in both museum and private collections. Tewhatewha that still have puhipuhi or taupuhi are particularly sought after.

In some districts the tewhatewha was also called the paiaka, after the Maori word for root. In H.W. Williams' *Dictionary of the Maori Language* it is also suggested that the name wahaika was given to a weapon of bone or wood that was identical to the tewhatewha, although this name does not appear to have been in widespread use for this particular weapon.

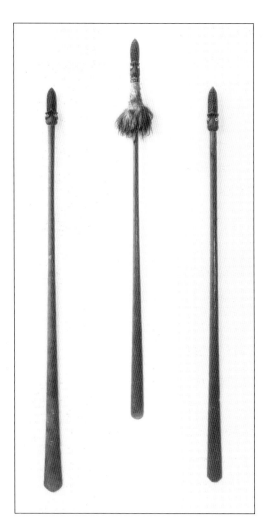

Intricately carved and beautifully balanced, the taiaha was the most common double-handed weapon employed by Maori at the time of European contact. Relatively few taiaha in collections have retained the feather and dog-hair decoration that adorns the central taiaha.

TAIAHA

The taiaha is perhaps the best known of all Maori weapons, being well represented in museum collections around the world. It was known by several names throughout Aotearoa, including maipi, hani, and the dialectal variation taieha in parts of the South Island. Another name that can probably be added to this list is turuhi, which is described in Williams' *Dictionary of the Maori Language* as being similar in appearance to the taiaha, and generally made from maire. A second description of the weapon states it was 'about 6ft. long, of which about 2ft. 6 in. formed a broad blade $2\frac{1}{2}$ in. wide and $\frac{3}{4}$ in. thick'.

Taiaha varied considerably in length, ranging from 1.2 to 1.9 m or more, with an 'average' length of 1.5 m. The weapon can be divided into three separate sections. The first was the rau, or long striking blade, which was fairly consistent in width, usually between 6 and 7 cm wide, although individual preference meant that some rau were as narrow as 4 cm. Next came the shaft, which was somewhat oval in cross-section. The third section was the proximal end, which featured two stylised upoko, or heads, carved back to back, with an arero or tongue extending out from the mouth in the Maori gesture of defiance. Generally, both sides of the arero were carved in an intricate pattern of curvilinear designs. The arero formed the extreme end of the weapon.

Usually both upoko carved above the arero featured a pair of eyes inlaid with discs of paua shell, although other shells were used occasionally, including oyster, pipi and koeo. Strong adhesives such as the gum from the tarata tree, and a flax gum, termed piaharakeke, were used to keep the discs in place. Some tribes called the gleaming eyes mata a ruru, after the eyes of the native owl, the ruru, while others believed that the taiaha's brightly shining eyes watched out for the enemy in hostile times. Occasionally, completed taiaha are found with two or more of the pupils left uncarved. The following excerpt from an article by C.S. Curtis in the *Journal of the Polynesian Society* explains this curious practice:

Several times in the past I have noticed this unusual feature of the eyes on the carved heads of the Maori weapons known as taiaha. The eyes did not have the usual inlay of paua shell. On one side of the weapon the eyes are shown without a pupil, but the pupils are shown on the eyes on the other side. When I first noticed this feature I assumed that the carving was unfinished. Since then I have mentioned the matter to a Maori friend and he informs me that this style was fairly common in the olden times. He states that the side of the weapon with the eye open was called mataara and the side with the eye closed matamoe.

Extremely tough, dense-grained hardwoods were usually used for making taiaha — timbers such as kanuka, manuka, akeake, ake rautangi, and black and white maire, which were valued for their resilience and relatively light weight. In a few rare examples whalebone was used. The construction sequence for wooden taiaha was basically the same as for the pouwhenua, with the addition of the carving of the head and tongue, which replaced the smoke- or fire-hardened point of the pouwhenua.

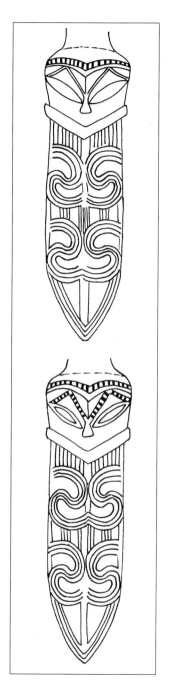

Top: Matamoe.
Bottom: Mataara.

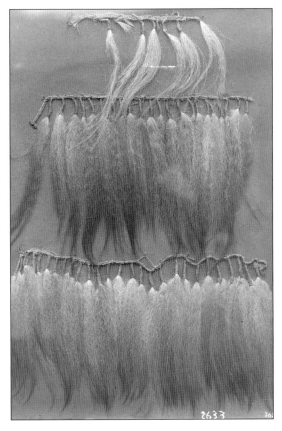

Many taiaha were adorned with red feathers and selected tufts of awe, or dog hair. A band of red kaka feathers, termed a tauri kura, was attached to the shaft of the taiaha, just below the carved head. In the best-made tauri kura, the feathers from beneath the wing of the kaka were intricately plaited into a length of woven fibre termed a puni. (An inferior process developed after the introduction of European cloth, using an imitation puni of red trade cloth. It was often necessary to bulk up this imitation puni by placing split raupo reeds under the cloth.) To complete the decoration of the taiaha, small tufts of long white hair, awe, from the Maori dog were sometimes attached under the red band of feathers. When adorned in this way the weapon was called a taiaha a kura, sometimes abbreviated to taiaha kura. When the weapon was not in use, tauri kura and taiaha a kura were

This image shows rows of awe (dog's hair) lashed to a broken cord, as they would look before being wrapped around the neck of a taiaha. Careful consideration must have been given when adorning any weapon with awe and feathers, so as not to upset the fine balance of the taiaha.
(Collection of the Museum of New Zealand Te Papa Tongarewa, C.906)

often protected by a sheath of parapara leaves.

Elaborately decorated taiaha such as these were often held by chiefs and other important men when they made speeches at tribal gatherings and other important hui. They were seen as an emblem of rank at such events, and to challenge a rangatira out of turn while he was holding such a weapon was to take your life in your hands.

The main use of the taiaha, however, was in battle. When advancing to engage an opponent a warrior often assumed a guard, with the taiaha held either vertically or slightly diagonally across the body, the blade uppermost and the arero facing the ground. During fighting, feints and passes were constantly employed, and experienced fighters continually repositioned themselves, hoping to detect an opening for a strike, all the while guarding themselves against an attack. A favoured ploy was to feint an attack on an enemy's torso or face with the tongue end of the taiaha then, while the opponent recoiled, whaka-koemi, or reverse arms, and strike at the top of his skull with the edge of the rau, or blade. When carried out with power and precision, such a blow could cave in the top of a skull and kill instantly.

One of the tricks young warriors were taught when learning to use the taiaha was to watch the big toe of their opponent's leading foot. It was believed that an impending attack could be anticipated as the enemy warrior gripped the ground with his toe, giving the defending warrior a fraction of a second in which to move or defend himself.

While the taiaha was designed to be an effective and efficient killing weapon, it wasn't only during war that it proved its worth. The following eyewitness accounts, both written well after the taiaha fell out of favour as a weapon of war, were published in Rotorua's *Daily Post* in March and April 1965 respectively.

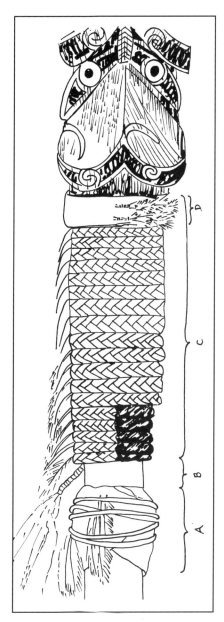

Line drawing showing detail of the taiaha a kura from *'Mick's Workbook: Drawings and notes on Maori artefacts and construction techniques. Book one'*, by Mick Pendergrast

A. Brown paper, overlaid with Kaka feathers held in place with rolled cord.
B. Tags of white dog hair tied in bundles that are joined together with single pair twining.
C. First layer of black rolled cord, overlaid with three-ply braid of fibre that holds kaka feathers and a few small brown feathers in place.
D. Cloth covered in small feathers (kaka and small brown and small blue/black feathers) sewn into place.

(Auckland Museum, AM 1353).

A selection of carving designs used on taiaha. Unfortunately, information relating to tribal origins is unavailable for these taiaha. Top row, L–R: AM 38788, AM 9, AM 34119. Bottom Row, L–R: AM 22491.2, AM 53489, AM 3790. (Auckland Museum)

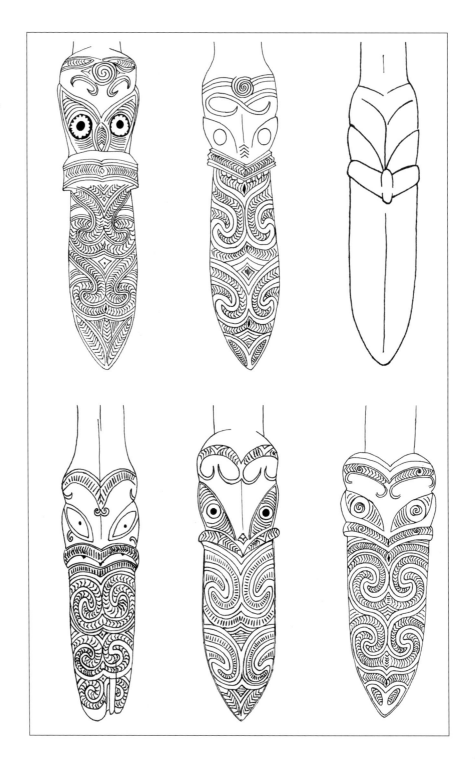

The old Maori weapon the taiaha can be deadly when wielded by an expert. This was proved in a taiaha and rifle and bayonet duel at a small arms weapon training school at Maadi, in the Middle East, in 1943. The school was an important centre in which men of the 8th Army were given an intensive training course in every infantry weapon, from revolvers to bayonets.

In this particular course there were Americans, Free French, English, New Zealanders and a sprinkling of Cypriot and Canadian troops. Most of them were junior officers, and the pace was on in readiness for Montgomery's all out 'push' at El Alamein.

In one of the bayonet fighting sessions, Major (then Lieut.) Don Stewart, Whakatane, remarked to his hard-bitten instructors: 'This is quite a weapon, I only know of one to beat it!' 'What's that?' asked the instructor. 'The Maori taiaha.' 'What the **** is that?' 'A fire-hardened wooden stave and fending spear,' replied Lieut. Stewart.

The derision and scorn this remark provoked, stung the young Whakatane man to the quick. As a result he offered to prove his point. Immediately bets were offered at great odds that the man with a Maori weapon would be dead within seconds against an expert with a rifle-mounted bayonet.

So it was arranged — the best bayonet exponent was to be matched against an expert armed only with the Maori taiaha. The event caused more than a stir when it was seen that Don was serious. Time and place were arranged. Money flowed.

The Maori champion was Lieut. Aubrey Rota, Matata, who had luckily brought one of the time-honoured weapons with him from New Zealand. There was a large and amused audience representing every section of the trainees at the fighting pitch. Rota was warned that he would have to take full risk of being wounded or worse, and that the incident was to be officially regarded as an exercise in the combat school. 'Accidents' were a fairly frequent occurrence in combat school exercises.

The young Maori was willing enough. There would be no holds barred on either side. Stripping off his tunic, he stood facing the grinning modern in much the same way as his forebears had faced the British redcoats a century ago. Rota was an expert.

The signal to start was given.

The soldier lunged and thrust in perfect precision, but each move was parried by the light-footed Maori who bided his time and stood on the defensive. The Pakeha grew cunning. Failing to penetrate the Maori's guard he feinted and tried to tempt an attack. But his opponent was too wise. There was rising anger in the actions and expression of the khaki warrior as thrust after thrust was tossed aside by the stout wooden weapon. Sometimes it was repelled with such violence that he was flung sideways.

To the accompanying cries of the somewhat dumbfounded onlookers the soldier gathered himself. Feinting smartly, he employed an old Army trick, crouched and charged directly at the Maori's midriff.

This was Rota's chance. Grasping his weapon firmly, he sidestepped, tipped aside the blind thrust, and caught the lunging figure a smart upper-cut in the stomach with the bladed end of the taiaha. In a flash he whirled the weapon about, to crash the business-end on top of his opponent's skull. Down he went, to be out of action for some days in the camp hospital — another regrettable accident from the small arms school.

The effect upon those present was profound. Money changed hands at great odds, as the jubilant minority collected. The story was repeated with almost unbelievable astonishment throughout the Middle East.

The second incident was recorded by Dr D.A. Bathgate. In this instance the contest took place in Bellamy's restaurant in the late 1800s, and was witnessed by numerous members of Parliament.

The witness of this remarkable duel was an old military friend of mine who one night found himself in Bellamy's with his colonel discussing the pros and cons of weapons of war in general. The old colonel was emphatic that nothing could touch the sword as a personal weapon. In the hands of a champion, he said, it could easily outclass any other weapon known.

The conversation was interrupted by a well-known Maori MP who politely remarked: 'Excuse me, sir, but I do not agree with you. Our Maori taiaha in the hands of an expert is superior to your Pakeha sword. Furthermore, if you wish it pick the best swordsman in your camp and I will match him with a Maori champion and prove my words. If you like we can arrange a duel here, and to prove my sincerity here is £100 to back my man.' The only concession asked by the MP was that he be given two weeks in which to produce his champion.

The colonel was taken back, but in the face of all the excitement, accepted the challenge. Drinks were served, and bets made. The date was fixed and the news spread.

In Newtown camp at the time there happened to be a prominent swordsman who had twice won the honours at Aldershot. He was a regimental sergeant-major, perfectly built and completely confident.

The interest was terrific. Bellamy's was crowded, not only with parliamentarians but with military and naval personnel. Odds were heavily against the chances of the Maori. The sergeant-major was there, smartly uniformed as a member of the Royal Horse Artillery. He cut a striking figure, calmly assured and fully confident of the superiority of his weapon. He made a tremendous impression.

The Maori MP turned up with his champion — a giant reputed to be 80 years old. There he stood in front of the select and fashionable Pakeha audience clad only in his flax piupiu. His head was adorned with a flowing mane of snow-white hair and the heavy tattoo marks on his wizened old face gave way to an unkept beard reaching almost to his waist. In his hands he clutched the weapon of his forefathers. It was a beautifully carved taiaha, smooth and polished till it shone. He cut a truly Homeric figure as he stood waiting for the signal to begin …

With the word 'engage' the Maori at once took the offensive. With a shout and a bound he made a sweeping blow at the legs of the soldier.

Keyed and alert, the officer neatly parried and evaded the action, bending slightly forward as he did so. The Maori's reaction was almost too swift for the eye to follow … Leaping upward and back, with a swift reversal of his weapon he struck a sharp upward blow with the spear-ended stave. It connected with the Pakeha's chin, lifted him clean off his feet and hurled him flat on his back to the floor.

There was dead silence as the spectators watched the Maori champion move over to the still and apparently lifeless figure. The Maori knelt beside the inert form, gently touching and talking softly.

It took a long time for the vanquished officer to come round, and when he did he had to be carried from the field of battle.

The duel had lasted exactly 30 seconds! Bets were settled, and Bellamy's buzzed as the drinks were consumed to the honour of the old Maori warrior.

PATU AND MERE

The most common patu and mere are the patu onewa, patu paraoa, mere pounamu, kotiate and wahaika. Elsdon Best also collected the names meremere, mata kautete (described in Williams' *Dictionary of the Maori Language* as 'a weapon of sharp flakes of flint lashed firmly to a wooden handle') and mira tuatini (which Best states was a generic term for a patu, although Williams describes it as a saw-like cutting instrument made by lashing strips of obsidian or shark's teeth to a wooden handle; tuatini is also the word for a whaler shark or seven-gilled shark). Although the term mere is commonly applied to all patu, it should really be restricted to those fashioned from pounamu.

Each of the patu and mere that are described here was a flat, single-handed weapon averaging some 36 cm in length. Primarily designed for thrusting, they all featured an elongated or oval blade with a sharp striking edge.

Several materials were used to make patu and mere, including volcanic stone, pounamu and sperm-whale bone. Some locations were famous for the abundance of suitable stone for weapons; the West Coast of the South Island, for example, was universally known for the pounamu that was to be found in various pockets there, while the headwaters of the Waipaoa River in the Poverty Bay district were well known to local iwi as a place where the dark stone termed uri was found.

Like all Maori weapons, being hand made ensured that no two patu or mere

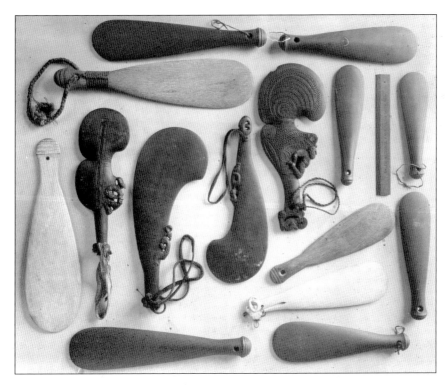

A magnificent selection of patu and mere. Of particular note is the lashing on the handle of the patu paraoa, second from top on the left, and the surface carving on the wahaika, second from left in the centre.

were exactly the same, usually differing both in size and weight. Typically, these short weapons were either held in the hand, secured with a dog-skin thong that passed around the wrist, or carried thrust into a woven war-belt.

Warriors using such weapons in battle relied heavily on quick footwork and agility. Some were so skilled that it was not unheard of for a man armed with a mere or patu and an old cape wrapped around his free arm to fight an enemy armed with a spear or taiaha. Typical strike zones for warriors fighting with mere and patu included the temple, the jaw and the ribs. In each case the leading edge of the weapon was used, rather than a downward clubbing action. This evolution away from the more typical downward blow common to other forms of club led to the emphasis on sharpening the front edge of the weapon, and to a lesser extent its sides. A secondary development was the introduction of a bored or chiselled hole to accommodate the wrist cord, which became necessary to stop a blood- or sweat-drenched hand slipping up the weapon during the thrust.

In giving names to particular weapons, Maori simply added the name of the material used in their manufacture. For example, a patu made from whalebone (paraoa) was termed a patu paraoa, while a mere made from pounamu was termed a mere pounamu.

Patu onewa

The patu onewa was perhaps the most common form of patu produced by the Maori. While it is widely exhibited in museums around the world, its fine finish and intrinsic beauty are often overlooked when compared with patu paraoa and mere pounamu. Patu onewa were made from a variety of fine-grained rocks, and the finish achieved is testament to the skill of the Maori in working with the most basic of materials and tools. Elsdon Best collected the following information about the names of stone used in making patu onewa:

The native names of the kinds of stone used for the manufacture of patu and toki were — kara, a basaltic rock; uri, a stone somewhat resembling the kara; onewa, a dark grey stone; tuapaka, a white or grey stone; kurutai, a dark (pango) stone; makahua (he mea whero) [brown or reddish]. The kara appears to be the stone known to the Ngati Hau tribe of Whanganui as kororariki. I have heard it stated that onewa is another name for koma, a light-coloured stone of which implements, &c., were made, but Williams gives them as two distinct kinds. Anyhow koma and onewa are both terms applied to toki or patu made of light coloured stone.

A finely finished patu onewa from the Te Papa collection, photographed in front and side profile.
(Collection of the Museum of New Zealand Te Papa Tongarewa, B.16921)

In construction, the craftsman's first step after selecting his stone was to begin shaping it into the desired profile by a process of flaking. The shape was then refined with hammer stones of quartz or other equally hard stone, by which time the patu would be the approximate shape of the finished article, being perhaps 2 mm larger all round.

Once the basic shape had been formed the edge of the blade was progressively fined down, initially with light hammer stones, then by grinding with sandstone. Next, a flat rim standing 4 or 5 mm high was ground along the edge of the blade. This was then carefully reduced further by grinding the blade's surface above and below the rim, leaving a flat edge of perhaps 1 mm. This edge was then left while the rest of the construction continued, in order to limit the possibility of damage occurring during the rest of the work.

Next, a hole was drilled in the handle for the tau, or dog skin cord, which was used to suspend the weapon from the warrior's wrist. The hole was started with a hammer stone, which formed a slight depression, before a drill called a tuiri or horete was introduced to finish the work. One version of the drill was formed around a stick shaft perhaps 30 cm long. Into one end of the shaft was inserted a stone drilling point chipped from a hard stone such as kiripaka, or quartz. Fastened at the mid-point of the shaft were two stones, one on either side, which helped keep the momentum of the drill going once it was in use. The rotation of the drill was accomplished with the help of two strings tied to the shaft above the stones. One string would be wrapped around the shaft, then pulled to make the drill turn; with the weight of the stones helping the drill turn, the second string was soon wrapped about the shaft. The strings were then continuously pulled one after the other, and the drill was kept in motion. The drilling, which was done from both sides of the handle, formed two cone-like holes which met in the middle.

A selection of patu onewa butts showing typical reke designs.
L –R: AM 849 — collected by Captain Gilbert Mair, location unknown; AM 48880 — collected from Peach Grove, Ahuahu, Great Mercury Island; AM 54453.2 — collected from Oruarangi, Hauraki Plains.
(Auckland Museum)

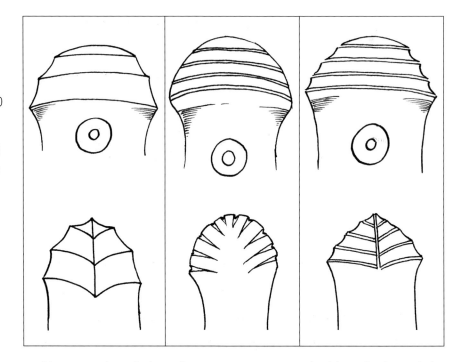

Next, a series of channels or grooves were etched into the butt of the handle. Although the grooves seem to have been purely cosmetic, a considerable amount of time and effort would have been needed to produce them. The number of grooves varied, with most patu having between three and seven. The final act for the craftsman was to return to the edge of the blade and sharpen it with a grindstone, working from the centre of the blade outwards.

A story attributed to Elsdon Best in W.J. Phillipps' book *Maori Life and Custom* gives an intriguing account of Maori commerce in Hawke's Bay:

A certain Maori chieftainess of high lineage who lived in the Hawke's Bay in the 1890s became financially embarrassed and called together some of the wise men of the tribe. Two men who were expert artificers were first summoned, and it was agreed that with the help of a grindstone and other tools, a large number of patu onewa would be

made and sold to unsuspecting Pakeha collectors. Helpers were
assembled and all was carried out according to plan. The patu onewa
are said to have been exact replicas of samples from the pre-pakeha
days. For a while the tribe and their chieftainess grew rich; but all good
things come to an end, and the market was soon saturated. How many
of these spurious patu onewa have found their way into museums we
cannot now say.

As with all Maori weapons, warriors learning to use patu onewa trained using predetermined moves that focused on coordinating both foot and arm movements. These exercises were designed specifically for individual weapons. In battle the patu onewa was generally used with a thrusting motion to attack the enemy's upper body, or to finish off a wounded enemy with a downward blow to the head using the reke, or butt.

This story, collected by Captain Gilbert Mair and published in Best's *Notes on the Art of War*, describes a fight between two warriors, one armed with a patu onewa, the other with a type of spear called a tokotoko:

A famous single combat was that fought out by Te Purewa, of Tuhoe,
and Te Waha-kai-kapua, of Te Arawa, on the bloody field of Puke-kai-
kāhu. Korotaha acted as second (piki) to Te Purewa, and Toko
performed a like office for Te Waha. Then was seen a Homeric combat
as these two giants strove together. Te Purewa fought with a patu
onewa, his opponent used a spear (tokotoko). The former, in warding
off a blow, had his weapon broken, the stump thereof alone remaining
in his hand. Waha then pierced him in the shoulder with his spear, felling
and pinning him to the earth. Korotaha strove to save his principal, but

was attacked by Waha's second, and thus had his hands full. Waha shortened his grip on his spear in order to drive it home, when Te Purewa, with a desperate effort, struck upwards at the temple of his foe and slew him with the stump of his patu.

PATU PARAOA

The patu paraoa, which was usually fashioned from sperm-whale bone (paraoa), was an exceptionally strong and well-balanced weapon. Particularly well-made patu paraoa, or those that had featured in events of tribal importance, were highly valued heirlooms that were passed from father to son, and over time attained a beautiful golden patina from generations of handling and use.

Generally symmetrical in form, the patu paraoa was a simple flat-bladed weapon that, like other weapons made by the Maori, varied in shape and size from example to example. Most patu paraoa averaged some 43 cm in length, and were perhaps 1.5–2 cm deep at the thickest point. There were, however, a small selection of patu paraoa that were considerably thinner in cross-section. Averaging no more than 2 or 3 mm in thickness, and slightly convex in profile, it seems that these patu were shaped from the crown of the sperm whale (see page 37, bottom image).

Like patu onewa, patu paraoa typically had a hole in the handle through which a cord, or tau, was threaded, long enough to wrap firmly around the warrior's wrist. The majority of the cord holes were made using the flint-tipped drill called a tuiri or horete, but some patu were finished with a rectangular hole for the tau, which was shaped with the help of a chisel. It is unclear why this is, but it is possible that this technique may have been left over from before the introduction of the drill.

Left
Unlike the majority of whalebone patu that have the more common etched grooves on the reke (as seen on the patu paraoa at the far right), this example has been finished with a finely carved figure.

Middle
Beautifully carved, this rare patu paraoa has a stylised human head carved into the blade.

Right
A typical example of the etched groove finish often used on patu paraoa.

Bottom
Side profile of the thinner patu paraoa cut and shaped from a portion of the skull of the whale.

Aesthetically pleasing patu paraoa (at least by twenty-first century standards) had intricately carved figures or well-carved grooves etched into the butt of the handle, similar to those seen in the best patu onewa, while lesser examples were finished with simple, often irregularly spaced horizontal grooves.

Patu paraoa were used both by warriors in war and by chiefs in speech making. In combat, the patu paraoa was used as a thrusting weapon rather than as a club, the most lethal strike being the straight jab, or tipi.

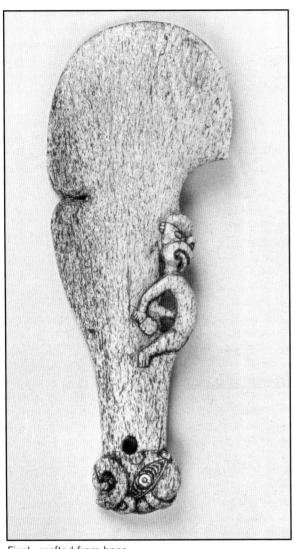

Finely crafted from bone, this wahaika carries both the carved notch and human figure above the handle. Of interest is the short handle: based on contemporary hand sizes, there is only enough room to wrap three fingers around it.

WAHAIKA

Although the shape of the wahaika (which literally means 'mouth of the fish') was quite distinctive when compared with other forms of patu, this was perhaps the least uniform of all patu and clubs, with many small variations in design. Wahaika were made of both whalebone and wood. As may be expected, a larger proportion of the wooden wahaika have surface carving on the blade when compared to those fashioned from bone. A further variation can be seen in a number of very thin, curved wahaika that were fashioned from the crown of the sperm whale's head.

The most striking features of the wahaika were the concave back and the peculiar notch that is carved into the edge of many surviving examples. A number of wahaika were also carved on the reke, or butt, and significant numbers also had small human or manaia figures carved above the handle.

Like other weapons, the wahaika was not only used in battle, but also in ceremony and speech making, where it was used by rangatira to accentuate the delivery of particular points.

The wahaika was primarily used for close-in fighting, being designed for thrusting rather than smashing down in a club-like manner. The favoured strike was made with the sharpened leading edge of the wahaika.

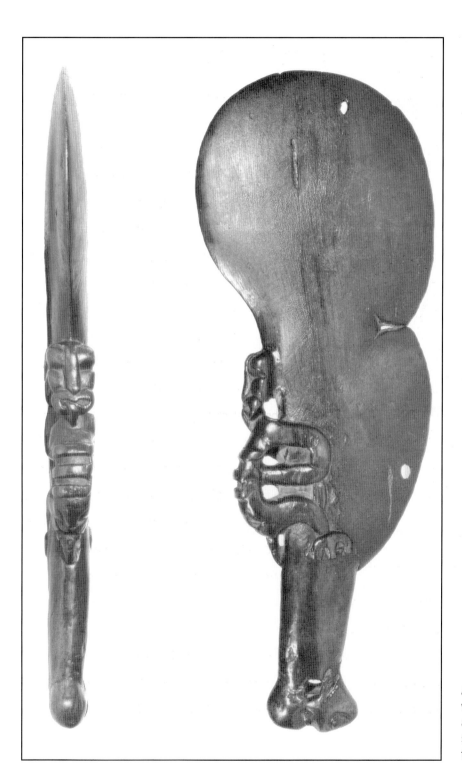

A fine example of the
wahaika, shown in side
and front profile, with a
square chiselled hole for
the wrist-cord.

MERE POUNAMU

The mere pounamu was perhaps the Maori's most revered weapon, due in part to the efforts required in its construction, as well as to the relative scarcity of pounamu in most areas of Aotearoa. While some mere were buried with their owners, many more were passed from generation to generation, gathering prestige through association with their famous owners. There are numerous examples recorded of mere that were given personal names and acquired particular virtues. Te Rauparaha's mere pounamu was one such; named Tuhiwai, it supposedly had the power to foretell the future, which it expressed by changing colour. The fame of such weapons was so widespread that prisoners sometimes asked to be killed by a specific mere.

Mere pounamu differed considerably in length and weight from example to example. A study of mere pounamu in museum collections suggests that most were between 24.5 and 43 cm long, with the average being 35.5 cm. These variations in length were undoubtedly determined by the dimensions of the raw material that was available.

Suitable boulders of pounamu were most commonly found on the South Island's West Coast, and were a valuable commodity for the people of that region. The rock itself was extremely hard and difficult to work with, and the production of a mere pounamu required much patience.

Well shaped from nephrite (greenstone), this mere pounamu has a woven pouch around the butt. Extremely rare in modern collections, such pouches probably assisted with gripping the weapon in the heat of battle. (Collection of the Museum of New Zealand Te Papa Tongarewa, F.792)

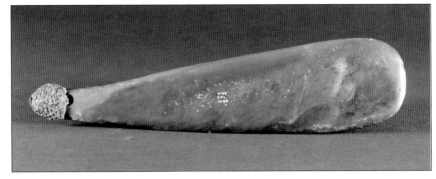

These three line drawings demonstrate the basic stance, defence and counterattack positions against a spear with a mere and puapua (any garment wrapped around an arm and used to deflect weapons). The use of puapua was reasonably common in battle, but for obvious reasons was limited to warriors deploying single-handed weapons. (Augustus Koch, 1891; Alexander Turnbull Library, Wellington, B-110-033-4, B-110-005-1, B-110-005-2)

The first step was to cut a suitable piece of pounamu from a boulder. This was achieved by making two parallel cuts across the surface of the boulder using a file of sandstone and a little sand and water. When the cuts were deep enough, the slab of pounamu was broken off and the tedious process of shaping the mere began.

Elsdon Best gives this explanation of how the pounamu was shaped:

The chipping implements used to roughly fashion an implement were simply pieces of hard quartz or similar stone (kiripaka) lashed securely to a handle. The piece of stone so used would be chipped to a rough point before being lashed to a handle. Both large and small hammers of this kind were used, the former for cleaving the block of stone to be operated upon, the latter for chipping into the required form. The fastening of the lashing was ingenious and secure (he mea kauiki te hitau).

When chipped into something like the desired size and shape, then began the long process of grinding. In various parts of New Zealand have been seen rocks, the surface of which are covered with grooves where the men of old were wont to perform the tedious grinding or rubbing process by which stone implements were smoothened and made symmetrical. It is indeed marvellous to note how well formed, symmetrical and true in outline are these stone implements of the Maori.

Once the mere had been shaped using the hoanga, or grinding stones, a tuiri (or horete) was used to drill the hole for the wrist-cord. Most mere pounamu were finished with a series of light, parallel grooves across the width of the butt. Some of the better-finished examples feature deeper, more regularly formed grooves, and these are likely to have been worked on over several generations.

When used in combat, the mere pounamu was basically employed as a thrusting weapon, and mainly used for attacking vital points of the head or body. Favoured strikes included a horizontal attack to the temple area, and an upward thrust under the ribs. When in use the mere was secured by a tau, or cord, wound tightly round the wrist, and if it was not in the warrior's hand it would be carried in his belt, usually secured against his back.

KOTIATE

The kotiate was a prized weapon on the battlefield, as well as being favoured by many chiefs during speech making. It was a curiously shaped weapon designed for close-quarter fighting, and noted for the carved notches on either side of the blade. According to popular tradition, these notches were used to entwine the intestines and other organs of the vanquished enemy — a tradition that is borne out somewhat by the translation of the word kotiate, which literally means 'to cut liver'.

The average length of kotiate was a little over 32 cm. They were usually fashioned from whalebone, and less frequently from a hardwood such as akeake or rautangi. Those made from wood were often intricately carved on the blades, while the carving on whalebone kotiate was usually reserved for the butt of the handle.

Like other mere and patu, the kotiate was finished with a small cord hole just above the butt of the handle, which was either drilled out, leaving the weapon with two cone-shaped holes meeting in the middle, or chiselled, leaving a rectangular hole. A cord was passed through the opening and wrapped around the warrior's hand to ensure the weapon wasn't lost in the heat of battle.

When using a kotiate against longer weapons such as a taiaha or pouwhenua, a warrior's main strategy was to parry and sidestep blows until he could get within arm's reach of his opponent, where the longer weapon lost its advantage to the point where it was almost non-effective. The warrior with the kotiate could then strike at his enemy, using its leading edge with a quick jab called a tipi. A favoured strike zone was the temple, where a twist of the wrist upon contact could crack open the skull. Once an enemy had been disabled, a downward strike with the reke, or butt, was often used to finish him off.

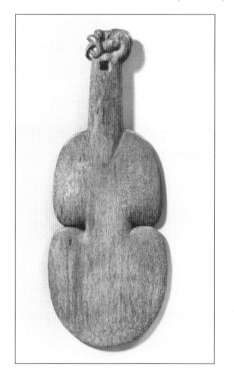

Left
This whalebone kotiate was probably fashioned prior to the introduction of European tools. None of the intricate carving often seen decorating later examples is evident, and the wrist-cord hole has been chiselled out, rather than drilled.

Right
Unusual for having its wrist-cord hole positioned side-on, this beautiful example of a kotiate clearly shows the close grain favoured for such weapons.

SPEARS

The spear was a common weapon in Aotearoa before the arrival of Europeans, being used in hand-to-hand combat, sometimes as a missile while approaching enemy forces, and for both defending and attacking the ramparts of fortified settlements. There appear to have been at least four distinct types of spear, although there is some difficulty in classifying them, as names sometimes differ from tribe to tribe, and in some cases different tribes use the same name to describe different spears.

The first class of spear included the longest, the huata. This was a long, slim spear with one end sharpened to a fine point while the other was often carved into a round knob. It was used almost exclusively to ward off attacks on pa, being too unwieldy to use in hand-to-hand combat. The tokotoko was similar in design to the huata, but a little shorter and so slightly less cumbersome. The next class encompassed the tao and koikoi, which were used primarily during man-to-man combat and were wielded with two hands. The points of such spears were usually hardened, often through the use of fire, giving them an almost steel-like quality. The third group encompassed throwing spears, such as the timata and tarerarera. Throwing spears were usually constructed either in one piece or with a lightly tied detachable point, which was designed to break off and remain embedded within an enemy. Throwing spears sometimes featured serrated points, which were very difficult to extract from the body. The fourth and by far most rare type of spear were those constructed with two or even three prongs. These spears, such as the reti and tararua, were carved from a solid piece of hardwood.

Early visitors to New Zealand noted that a popular pastime was a test of skill that involved avoiding flying spears. Warriors would attempt to deflect spears thrown at them by their comrades with the help of a short stick. This sport was also reportedly used as a means of settling arguments among kinsmen. The Frenchman Jean Roux noted in his journal that he often watched Maori '… throw darts, lances, assegais etc. at one another, warding them off with singular skill …'

Such skills were learnt from boyhood, in games that involved throwing and avoiding light raupo and toetoe stalks. Later, disagreements between men of the same village were often settled in like manner, with the transgressor set the task of avoiding, parrying or catching spears thrown by the injured party.

HUATA

The huata was the Maori's longest fighting spear, ranging in length from 4.8 to 7.5 m. It was pointed at one end, and sometimes had a fist-sized round knob, called a pureke or a purori, on the butt end. The huata was principally used when defending pa, although there are accounts of it being used by attacking forces, as described below. As with other spears, the thickest part of the shaft of the huata did not exceed much more than 2.5 cm in diameter. It was sometimes decorated with hair from the tail of the native dog, which was tied to the shaft above the purori.

Huata were most often fashioned from strong tawa or rimu. Large, straight-grained, mature trees were preferred because they split relatively easily and produced a true, clean wood to work with. Craftsmen used finely honed toki, or adzes, to reduce the diameter of the timber to within a couple of millimetres of the desired size, before scraping the huata along its length with stone flakes or shells to remove any final bumps or wobbles. The shaping process was then finished with an all-over rub using fine-grained sandstone, which was sometimes followed by rubbing the spear along the trunk of a tree fern to give it a particularly smooth finish. Finally, the point of the huata was hardened by fire, and like all other wooden weapons it was given a coating of shark oil to help prevent it splitting. Continuing maintenance included coatings of hinu tangata, or human fat.

The following account of long spear making, written by William Colenso, describes an alternative and somewhat elaborate technique used by members of the Urewera tribe, as witnessed at Ruatahuna in 1841.

First, a straight, tall, and sound tawa-tree was selected in the forest. This was felled with their stone axes. Its head and branches having been lopped off, it was dragged out into the open ground, and split down the middle into two halves. If it split easily and straight, then it would probably serve for two spears, if each half turned out well in the working. The next thing was to prepare a long raised bed of hard tramped and beaten clay, 35ft.–40ft. long — longer than the intended spear — the surface to be made quite regular and smooth. On to this clay bed the half of the tawa-tree was dragged, and carefully adzed down by degrees, and at various times, to the required size and thickness of the spear. It was not constantly worked, but it was continually being turned and fixed by pegs in the ground, to keep it lest it should warp and so become crooked. It took a considerable time —

about two years — to finish a spear. The last operation was that of scraping with a broken shell or fragment of obsidian, and rubbing smooth with pumice-stone.

Colenso goes on to explain that the huata, '… was used in defending their forts and stockades, being thrust through the palisades at close quarters against the legs and bodies of the invaders'. This practice was generally accomplished by the warrior balancing his huata on the horizontal posts, or huahua, that bound the palisades together. He was then free to manoeuvre the spear in relative safety.

Some early visitors, such as John L. Nicholas, who accompanied Samuel Marsden to New Zealand in 1814–15, reported that huata were fixed with detachable barbed points of bone, but it is almost certain that these were bird-hunting spears, which were finished thus to stop speared birds escaping.

When not required, huata and other spears were either kept together in a tribal weapons storehouse, or suspended from the roof of a whare where they could be grabbed quickly in an emergency. Spears kept in roof spaces were said to turn black from the soot of the fires below, and the constant exposure to the cooling smoke also helped harden them.

Although, as mentioned above, huata were usually used when defending pa, there are examples of their use in attacking pa. One such tradition recalls the fall of Oputara pa at Whirinaki, where Tuhoe warriors used bundles of burning ferns tied to their huata to set fire to the houses and defences within. With smoke and fire everywhere, the inhabitants were forced to abandon their stronghold. Huata were also used against Ngati Huri during fighting at Papakai pa; in that instance, the huata were used once the pa had been breached to slay people hiding within their houses.

Only a few terms relating to the use of the huata are now remembered. 'Awhipapa' was to advance in a low, stooping attitude while dragging the huata behind; 'amo' was to carry the huata on the shoulder, while 'ahei' was to hold the spear against the collarbone as a guard. The meanings of other terms, such as pitongitongi, hiki, whitiapu and kuku-a-mata, have been lost.

TOKOTOKO

The tokotoko was basically a shortened version of the huata. It had a fire- or smoke-hardened point at one end, and more often than not a pureke or purori (rounded knob) on the butt end. Unlike the huata, however, the tokotoko was short enough to be used in hand-to-hand combat if necessary. Typically made from manuka, the tokotoko was usually 3–3.6 m long.

The names of a few striking moves and guard positions for the tokotoko have been remembered: taki-whenua, ahei, whitiapu, kotuku and hiki. It is said that the matia was a spear that closely resembled the tokotoko, but no description of this weapon has been found, and it is possible that this was the name given to the tokotoko by one particular tribe.

The following story from Elsdon Best's *Notes on the Art of War* is one of the few narratives in which the tokotoko is mentioned in action.

Te Ranga of Ngati-Mahanga of Te Whaiti called upon his spearmen to raid the realm of Nga-Potiki. They marched to the head of the Wairau river and slew Parahaki, whose wife Mihi said to Te Ranga, 'He aha koia te mate noa ake ai, kei te ora nga toetoe tahae a Mihi-ki-te-kapua' — 'What does his death matter, the toetoe tahae of Mihi still lives', i.e. her son Whitiaua was still living and would avenge the death of his father.

Toetoe tahae is a coarse grass, the leaves of which have serrated edges which cut the hands deeply if carelessly handled. This reference meant that the Parahaki family was a dangerous one to interfere with. Mihi escaped and fled towards Rua-tahuna, ever wailing for her dead. Now her son Whitiaua was, at the same time, travelling to his mother's place, and sat down by the wayside to rest. Ere long he heard the voice of his mother as she wailed for her dead husband. Whitiaua knew that death had overtaken his people. But he did not move. As the mother came up the trail, he asked, 'Whose is the deed?' Mihi answered, 'Te Ranga of Ngati-Mahanga.' Whitiaua said, 'Proceed on your way.'

Before the moon had changed, Whitiaua had collected the warriors of Nga-Potiki, and advanced on Te Whaiti, where they surprised Te Ranga and his people at Huki-tawa. Many of Ngati-Mahanga were slain. Te Ranga fled, pursued by Whitiaua, who, after a long chase, felled his enemy with a thrust of his tokotoko, saying, 'How indeed may you escape from the toetoe tahae of Mihi-ki-te-kapua.'

Drawn by John Webber on Cook's third voyage to New Zealand in 1777, this beach scene at Queen Charlotte Sound, Marlborough, gives a rare early depiction of what is probably a tokotoko in the hands of one of the two warriors standing on the beach, centre left.
(John Webber; Alexander Turnbull Library, Wellington, B-098-015)

TAO

The most commonly used spear was the tao. Fashioned primarily using toki and quartz scrapers, in much the same way as the timata (described on page 54), the tao averaged between 2.1 and 2.7 m in length. One end was quite blunt, while the other was sharpened; like other spears, the sharpened point was usually hardened by fire. The kawau, or shaft, was thickest near the blunt end, being perhaps 2.5 cm in diameter at its widest point. Manuka, maire and akeake were the most favoured hardwoods for fashioning tao, although occasionally tanekaha, rimu and black hinau were used, the choice depending on the availability of suitable resources and personal preference. In a few rare examples, a simple carving is evident on the shaft of the tao.

Although by and large a very basic weapon, and to some extent mass-produced, the tao was nonetheless lethal in the hands of a quick and nimble warrior. Indeed, the tao is said to have often been the weapon of choice when individuals sought utu for any wrongdoing, real or perceived. It is said that many a fiery encounter was witnessed, but that a fight to the death was seldom seen, since it was widely understood that combat would stop at the first flesh wound.

Possibly because chiefs and famous warriors tended to use taiaha, pouwhenua or tewhatewha rather than the basic tao, few traditions have been recorded about its use. One notable exception is found in the story of Kupe, and his discovery and subsequent circumnavigation of Aotearoa. In this, Kupe is said to have thrown his taonui, or large tao, to the South Island while still astride the North Island. The spot where his spear landed, in the vicinity of Jacksons Head, Queen Charlotte Sound, was subsequently named 'Te Taonui-a-Kupe'.

Another story, this time recorded in Judge Maning's book *Old New*

This series of four images by Augustus Koch shows what is supposedly a typical fight sequence for warriors using tao. It seems probable that the spears used in this instance, despite being termed tao by the illustrator, were in fact double-pointed koikoi. As illustrated, the spears are used primarily to strike and parry with two-handed grips rather than to lunge forward with single-handed attacks.
(Augustus Koch, 1891; Alexander Turnbull Library, Wellington, B-110-010-1–4)

Zealand, tells of an old warrior passing on the fighting skills required by those using the tao to the young men of his village.

In the hot days of summer, when his blood I suppose got a little warm,

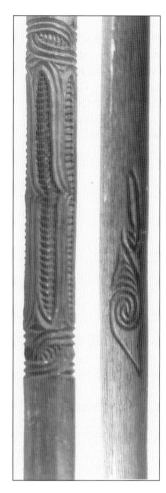

Detail of carving from the shafts of two tao.

he would sometimes become talkative, and recount the exploits of his youth. As he warmed to his subject he would seize his spear [tao], and go through all the incidents of some famous combat, repeating every thrust, blow, and parry as they actually occurred, and going through as much exertion as if he was really and truly fighting for his life. He used to go through these pantomimic labours as a duty whenever he had an assemblage of the young men of the tribe around him, to whom, as well as to myself, he was most anxious to communicate that which he considered the most valuable of all knowledge, a correct idea of the uses of the spear, a weapon he really used in a most graceful and scientific manner; but he would ignore the fact that 'Young New Zealand' had laid down the weapon for ever, and already matured a new system of warfare adapted to their new weapons, and only listened to his lectures out of respect to himself and not for his science.

KOIKOI

The koikoi was a relatively short, slim spear between 1.8 and 2.4 m in length, of similar construction to the tao but distinguished by being sharpened at both ends, rather than at just one end. Manuka or a similar hardwood was the usual wood of choice when making koikoi.

While the koikoi was probably relatively common, few examples have survived in museums or collections. This is probably because of its rather plain appearance and the fact that it was not usually carved.

Little has been recorded about the use or construction of the koikoi, but it is likely that it was made in much the same way as the timata. In use, blocking moves and thrusts may have been similar to those used with the taiaha and pouwhenua.

Maori haka group at Ngaruawahia with koikoi during the welcome for competitors to the 1950 British Empire Games. (New Zealand Free Lance Collection; Alexander Turnbull Library, Wellington, F-30847-1/2)

TIMATA

The timata was another spear of medium length, varying between 1.8 and 2.4 m. It appears to have been used both as a throwing spear and as a hand-held spear. Williams' *Dictionary of the Maori Language* states unequivocally that the timata was a dart or short throwing spear, while early statements collected from Ngati Awa claim that the timata was a thrusting spear 2.4–2.7 m long. It seems likely that while some tribes concentrated on honing its use in hand-to-hand combat, others preferred to use it as a throwing spear.

The following comments were recorded in the southern South Island region of Murihiku by Herries Beattie early in the twentieth century. Beattie touches on the use of throwing spears in the region, and an early name for such spears.

Nearly all my informants seemed to have heard of the throwing of spears in the old days of warfare. It would be strange indeed if they had not for spear throwing is mentioned at least seven times, if not more,

A Feast at Mata-ta. This beautifully detailed lithograph depicts a timata being thrown during a powhiri (welcoming ceremony) for a friendly tribe visiting a pa in the vicinity of Matata. (George French Angas, 1847; Alexander Turnbull Library, Wellington, PUBL-0014-36)

in the South Island record of fighting; four times recording the deaths of the chiefs at whom the spears were thrown, twice describing spear throwing contests and once relating the timely evasion and escape of the chief at whom the missile was hurled. The spear that was thrown is usually called timata nowadays but the collector was told that this word really applies to the act of throwing, the correct name for the spear being toro.

Beattie was told that toro were usually adzed down from hardwoods such as manuka, maire or akeake. Once an appropriate hardwood had been chosen, the first step was to singe off the bark with fire. Next the rod was adzed down in diameter, then scraped smooth with a type of quartz called kohairaki. Finally the mata, or point, was shaped, then hardened in a fire. Beattie's informant continued: 'When hardening spear points the manuka would sometimes make a whistling noise, called toi, and this was regarded as a very good omen for the user of that spear.' He concluded by stating that this hardening of the manuka by fire, if done properly, could make the point 'like steel'.

There are many examples of spears being thrown in Maori tradition, as well as eyewitness accounts recorded by early European visitors such as Captain Cook and Marion du Fresne. One well-known example is the death of the famous Ngai Tahu chief Tutekawa. Tradition has it that prior to leaving the Wellington province to settle in the South Island, Tutekawa was involved in a skirmish and killed several people, including two sisters named Tuarawhati and Hinekaitaki. The women's tribe planned to revenge their deaths, and some time later managed to trap Tutekawa within the fortified village of Waikakahi, near Lake Ellesmere. According to the story Whakuku, a brother of the murdered women, located Tutekawa's whare and, throwing his timata through the open window, killed his enemy.

Another story comes from Ngati Kahungunu of Te Wairoa. During the siege of the Ngati Kahungunu pa at Lake Whakaki by Waikato warriors, a Waikato chief called upon the famed Kahungunu warrior Te Rito-o-te-rangi to show off his spear-throwing abilities. The story says that Te Rito pointed out a man some two hundred yards away, indicating that he would be a fair target. As the Waikato chief turned to look at the man, Te Rito let fly with his timata. But instead of throwing the timata at the distant warrior, Te Rito aimed it at the inquisitive chief, who stood some fifty yards away. It is recorded that the hapless chief died on the spot, the spear passing clean through him. Not surprisingly, the Waikato warriors never again asked to see Te Rito-o-te-rangi's skill with the timata.

Spears such as the timata are also recorded as having been used during taki, or the ritual challenging of visitors, in the earliest days of Maori and European contact. When warriors approached Captain Cook and his men to challenge them, a single spear was invariably thrown. One such instance occurred near the mouth of the Turanganui River, where one of the challenging party was shot through the heart as he prepared to throw his spear.

TETE

The term tete was used to denote any spear with a detachable point. In combat tete were used primarily as a stabbing spear, although they were occasionally also used for throwing.

After being constructed in the same way as the timata, the tete was finished with a grooved notch in its forward end, into which the detachable point or barb was secured. The point, called a matarere, was variously made of wood (mapara, maire or occasionally katote, the hard black wood found in the trunks of tree ferns), human bone, whalebone (tete paraoa) or the spine of a ray (tete whai). The matarere was usually held in place by a two-ply cord that was tightly wrapped around the end of the tete.

While the occasional tete featured a small amount of surface carving, it was more usual for them to be ornamented with tufts of dog's hair (awe) tied about the lashings.

If the matarere was barbed, the spear might be called a kaniwha, while some tribes called any spear with a point fashioned from the tail of a stingray a tara-whai. All such barbed matarere are said to have inflicted severe wounds, and they were notoriously difficult to remove.

Below left
Close-up image of a detachable point lashed into the groove at the end of a tete. In this instance the original point has been replaced with iron, but the lashing and use of dog-hair tassels would have been identical for wooden or bone points.
(Auckland Museum, AM 2932)

Below Right
Details of a rare carving on the shaft of the tete.
(Auckland Museum, AM 2932)

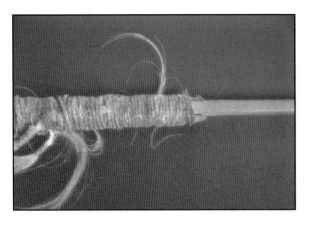

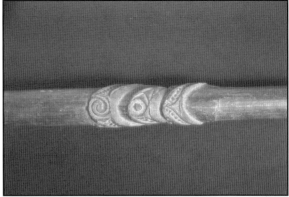

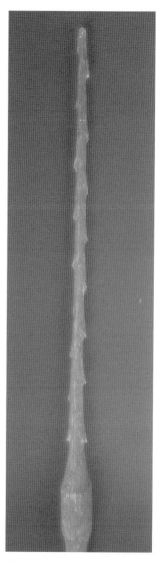

TARERARERA

The tarerarera was reportedly one of the most feared spears possessed by the Maori. Individual tarerarera varied in length, averaging about 3 m from point to butt. Reports collected by early visitors to New Zealand suggest that while the tarerarera was very similar in appearance to the timata, it had one major difference — the end of the tarerarera was barbed. The end was also often designed to break off on contact, leaving several centimetres of wood embedded in the body of the enemy. So effective was the design that it was next to impossible to effect a straightforward extraction of the broken section.

There are a number of eyewitness accounts that recall the use of the tarerarera in combat. The French, under Marion du Fresne, were on the receiving end of the tarerarera during their stay in the Bay of Islands in 1772. When the French attacked a pa the defenders hurled spears at them from behind the relative safety of palisades, wounding one sailor in the ribs and another in the thigh.

During his first visit to New Zealand James Cook noted that '... the darts are 10 or 12 feet long, and are made of hard wood and are barb'd at one end. They handle all their arms with great Agility particularly their long pikes or lances, against which we have no weapon that is an equal match except a loaded musquet [sic].'

John Rutherford, a probable ship deserter who lived among the Maori between approximately 1820 and 1825, witnessed a spear being cut from the thigh of a warrior. A simple shell was used to cut out the wooden end, which had broken off, and the resulting wound was said to be big enough to fit a teacup into.

In his writings Elsdon Best tells of the Tuhoe warrior Korokai, who was killed by his own tarerarera during fighting with Ngati Kahungunu:

Detail of the barbed prong of a tarerarera. Once impaled in the body of an enemy the sharp barbs would make the spear both difficult and extremely painful to remove. In this example the ravages of time have blunted the barbs. (Auckland Museum AM 35123)

57

This lithograph by Louis Auguste de Sainson depicts a group of Maori performing a haka in 1827 on board the French ship *Astrolabe*, captained by Dumont d'Urville. One of their companions, at extreme right, is in possession of a tarerarera. (Louis Auguste de Sainson, 1833; Alexander Turnbull Library, Wellington, B-052-021)

Should it so happen that the spear broke not at the notch when cast, it would be then taken and utilised by the enemy. During the battle of Puraho-tangihia, one Korokai, of Tuhoe, repeated the hoa invocation over one of these spears and cast it at the enemy. But he must have offended the gods, inasmuch as the hoa did not act properly on this occasion, the spear missing the mark, and the head thereof remaining intact. It was seized by Tama-i-runa, of Ngati-Kahungunu, who cast it back, slaying Korokai.

TARARUA AND RETI

The tararua, or tara waharua as it was sometimes called, and the reti are unique among Maori spears in that they had two prongs at one end, rather than the usual single point. Each spear was meticulously crafted from a solid piece of wood, usually the dense akeake or ubiquitous manuka. In examples where the pair of prongs were notched, the spear would be called a reti.

Very little other information seems to have been recorded concerning either of these spears, and only a handful survive in collections around the world.

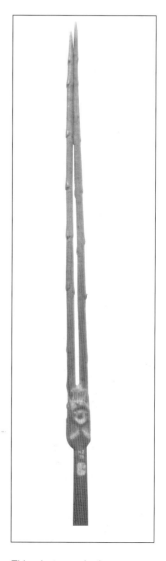

This photograph shows one of the few surviving examples of a reti. It was collected by Captain James Cook, possibly from the Whanganui district. The notches on the two prongs, although now quite worn, are clearly evident.

OTHER WEAPONS

HOEROA

The hoeroa is something of an enigma, the exact function of which is still keenly debated. Not all experts are convinced that the hoeroa was in fact designed as a weapon, despite a small number of stories that seem to support this view. A number of authorities suggest that the hoeroa was used primarily as a staff of chiefly authority, and only used as a weapon in emergencies. This confusion exists mainly because there are so few surviving reliable descriptions of the hoeroa in use from pre-European days.

The hoeroa was also known by the names tatu paraoa, huri taniwha, and paraoa-roa, and it was usually fashioned from the jawbone of the sperm whale. Hoeroa averaged between 1.5 and 1.8 m in length and 5 to 8 cm in width, and had a curved profile. The top of the hoeroa was often finely carved, while a number of examples also had a small amount of surface carving halfway down their length. On special occasions hoeroa were sometimes adorned with feathers and awe, tufts of dog hair. Elsdon Best described seeing a hoeroa decorated with either pigeon or hawk feathers and long white dog's hair, both of which were secured to the top end of the weapon.

Even among those who claim that the hoeroa was a weapon, there are conflicting statements as to its role. Some accounts claim it was employed as a striking weapon, being held in the hands in much the same way as the taiaha. Others suggest that the hoeroa was spun above the head and then thrown through the air after a fleeing opponent, the owner holding a cord that was used to retrieve the weapon. The great Maori scholar Te Rangi Hiroa stated emphatically that the hoeroa was used 'mainly in pursuit, and thrown with an underhand movement to connect the edge with the spine or loins of a fleeing enemy. Without slackening speed, the pursuer hauled on the cord which was attached to his left wrist and the weapon came to hand without delaying the hunt for another victim.'

The following excerpt from Elsdon Best's *Notes on the Art of War* describes a fight between two adversaries, one armed with a taiaha, the other with a hoeroa.

Whakapa called on his fighting men and marched on Puketapu (at Te Teko). They were seen by Nga-Maihi, who marched out of the fort to meet the enemy. Whakapa then advanced from his men, who were in column formation (matua), and Te Au-whiowhio left the Nga-Maihi column. The two met in the open space between the two forces, and there engaged in single combat (tau mataki tahi). Whakapa struck rapidly at Te Au with his taiaha, but the latter warded off the blow with his hoeroa and then, with guard and point, thrust the thin blade of his weapon through the ribs of his adversary before the latter could recover himself. So fell Whakapa of Te Pahipoto.

Above left
Detail of the design carved into the top of a hoeroa collected by Dumont d'Urville, probably during his first visit to New Zealand in 1827.

Above right
This hoeroa from the Te Papa collection has been finished with a small amount of carving halfway down the blade. It also has a small hole just below the primary carving for the attachment of feathers.
(Collection of the Museum of New Zealand Te Papa Tongarewa, B.18547)

OKA

The oka was a rare, dagger-like weapon that was usually fashioned from either wood or bone. Published stories of the pre-European use of oka are non-existent, and it is possible that the oka was in fact a copy of a European dagger, developed soon after initial contact. Evidence suggests that it was not used to any great extent, and that it really featured only in districts from the Waikato north. Little if any information regarding its use has been gathered from the southern end of the North Island or the South Island.

In perhaps the best-known account of the use of an oka, Colonel Walter Edward Gudgeon tells of an old chief named Nga Tokowaru who was in the habit of carrying a bone dagger concealed in his war-belt. After a desperate battle in which he was captured, Nga Tokowaru was taken before the chief of the victors, one Te Patu. Taking his last opportunity to save his descendants from eternal shame and degradation, he drew his hidden dagger and stabbed the unsuspecting Te Patu. He then quickly smeared the chief's tapu blood over his own head and body, knowing that his captors would not eat his flesh once it had been covered with the chiefly blood of Te Patu.

Waikato tribes called their version of the dagger 'tete', suggesting that they may have simply employed the removable points of spears as daggers, since these were known by the same name.

PATUKI

The patuki is counted among the most rare of Maori weapons in modern-day collections. It is the only known single-handed weapon that was designed to be used with a downward clubbing stroke, rather than the thrusting motion used with other single-handed weapons such as the mere pounamu and patu paraoa.

Approximately the same length as the mere and patu, the patuki was usually made from wood, although whalebone examples do exist. Patuki were triangular or rectangular in cross-section, and had slightly convex sides. They were often beautifully finished, with the finest of designs intricately carved into their surface.

Like most other mere and patu, the patuki usually had a small hole in the handle for a wrist-cord to pass through. It seems that in most cases the hole was rectangular, having been formed with small chisels.

Intricately carved and inlayed with paua, this beautiful whalebone patuki was collected by Captain Gilbert Mair and added to the Auckland Museum collection in 1890. (Auckland Museum, AM 91E)

KOPERE

The kopere (also called pere, and possibly tarerarera by Tuhoe) was in its most basic form a rough, undressed manuka stick, improved only with the sharpening of one end into a point. In most regions the dart seems to have averaged between 60 and 90 cm in length, although Tuhoe examples, as described below, were as long as 2.7 m. In most cases the dart was propelled by means of a whip called a kotaha. In more elaborate examples a notch was cut some 5 cm behind the point, and in those made by the ancestors of the Waiapu people short points of kaka ponga, the hard fibres of the tree fern, were lashed to the shafts. In both these cases a short section of the projectile was intended to break off on impact.

Over the years a lot of doubt has been expressed about the authenticity of the kopere as a traditional Maori weapon, and indeed whether the Maori used

This fine example of an elaborately carved kotaha has unfortunately had its shaft cut short. Typical kotaha, as illustrated on page 67, averaged some 152 cm.

whip-thrown spears at all. The best support for the idea of the kopere as a traditional weapon comes from eyewitness accounts recorded in the journals and diaries of French sailors who visited the Bay of Islands in 1772, under the command of Marion du Fresne and his lieutenant Julien Crozet.

The French had plenty of opportunities to witness Maori weapons during their stay of some 33 days in the Bay of Islands. When a group of Frenchmen were shown around the armoury of a pa they were visiting, Crozet immediately recognised a number of kotaha among the piles of spears and clubs. The whips were described as sticks 'furnished at one extremity with knotted cord for throwing darts in the same way as we throw stones with slings'. A more threatening run-in with the kopere came about later during the Frenchmen's stay in the Bay of Islands. During the attack on Paeroa pa (led by Crozet to avenge the murder of du Fresne), two warriors, referred to as chiefs by the French, were seen to hurl darts at the French using whip-slings. Although the French considered the kopere to be an inaccurate weapon, they did acknowledge that it was able to be thrown relatively long distances with the aid of the kotaha.

A similar weapon to the kopere, called a tarerarera (not to be confused with the spear of the same name), was used by Tuhoe warriors. The following excerpt from Elsdon Best's *Notes on the Art of War* describes the Tuhoe version of the weapon.

> The *tarerarera* were, as stated, not finished or carefully made spears, but simply a rough throwing spear. Manuka was the favourite wood — small, straight saplings of about one and a quarter inches in diameter, and some nine feet in length. These were trimmed of branchlets and the scaly outer bark, the butt end was sharpened to a point, the same being hardened by fire. At about six inches back from the point, i.e., where the tapering off (*koekoeko*) commenced, a deep ring notch (*tokari*) was made, almost severing the head of the spear. On striking anywhere, the impact caused the head of the spear to break off at the ring notch, thus in striking the human body the head would remain buried in the body, causing a wound from which recovery was extremely doubtful.
>
> I have heard it stated that these casting spears were sometimes pointed with *katote* (*kaka ponga*), the hard, black fibres of the *kaponga*, or 'fern tree' of the colonists, which is of a poisonous nature. Also that this spear was sometimes thrown by hand, as in the siege of a *pa* (fort). Some natives state that *tuku whakarere* was applied to such a rough spear thrown by hand, and *tarerarera* to the one thrown by means of a whip.

Best continued by explaining that when used, the butt end of the *tarerarera* was stuck fast in the ground, and the cord of the whip was hitched round the shaft of the spear. A vigorous swing of the whip saw the spear depart, with 'a greater or lesser degree of accuracy', towards the enemy. Informants described old-time scenes where the flight of *tarerarera* 'resembled a rain storm'. The only effective protection against these fearsome spears was a wet pauku, or pukupuku cloak. Best also noted that tarerarera were sometimes termed manuka, from the wood of which they were formed, and that the term kopere was also used among some tribes to denote the whip.

The final word on the kopere comes from Gilbert Mair, a surveyor, interpreter and soldier who was in close contact with Maori from the 1860s on.

Over twenty years ago I placed a kotaha and several darts (or kopere) in the Auckland Museum.

The kotaha, or staff, is a stout piece of manuka about an inch and a-quarter thick and about four feet long, the lower end tapering and carved with a hole, to which is attached a stout bit of flax cord, terminating in a round, flattish knot; the upper end usually was carved to resemble a human hand. The kopere, three feet six inches to four feet, is made of the hardest wood heavily barbed, so that withdrawing it from a wound would be impossible; the end is a very sharp point, hardened in the fire. When used, the dart is stuck loosely in the earth at an angle and direction calculated to strike the object aimed at. The thong is then wrapped round the dart once, just below a small ridge, so that when propelled, the string frees itself in its forward flight.

The Ngati-whakaue, of Ohinemutu, Rotorua, were famous for their skill in using this primitive weapon, and on one historic occasion, while attacking Puhirua pa over a hundred years ago, one of the warriors, Te Umanui by name, hurled his weapon with so true an aim as to transfix a chief in the pa through the heart. The spot where Umanui threw it from was pointed out to me; it is a small eminence some eighty yards inland of the pa.

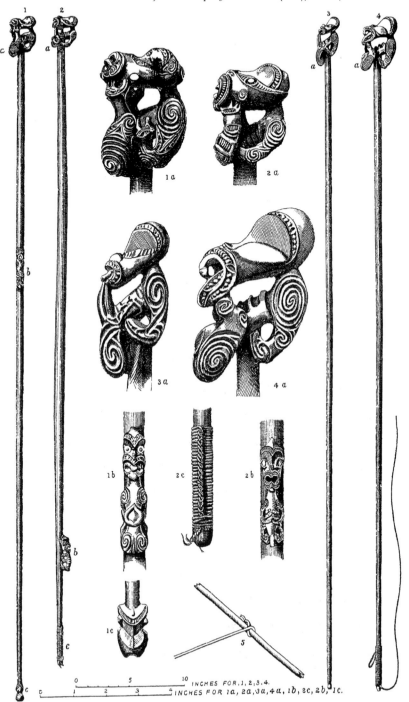

Journal of the Anthropological Institute (N.S.), Vol. II, Plate XXXIV.

NEW ZEALAND KOTAHAS OR WHIP-SLINGS.
FOR THROWING DARTS: IN THE BRITISH MUSEUM.

New Zealand Kotahas or Whip-slings. For throwing darts: In the British Museum. This fine illustration from the *Journal of the Anthropological Institute* shows a selection of kotaha from the British Museum collection. Of particular interest are the fine carvings seen on two of the shafts, and the lashing detailed in figure 2C.

GLOSSARY

arero	tongue of a taiaha
awe	dog's hair used for decorating weapons
hoanga	grinding stones
horete	a drill used on stone; also called tuiri
hui	gathering, meeting
kaka	a New Zealand parrot
kawau	the shaft of a spear
kereru	wood pigeon
kotaha	a whip or sling
manaia	carved bird-like figures
mata	point of a spear
matarere	detachable point of a spear
pa	fortified village
paraoa	whalebone
piupiu	flax skirt
pounamu	nephrite, greenstone
puhipuhi	the bunches of feathers sometimes attached to tewhatewha and other weapons; also called taupuhi
pureke	the rounded knob on the butt end of the tokotoko or huata
purori	the fist-sized round knob on the butt end of huata or tokotoko

rangatira	chief
rapa	the broad, quater-round head at the striking end of a tewhatewha
rau	blade
raupo	bulrush
reke	the butt of a patu or mere
tapu	sacred, taboo
tau	the wrist-cord attached to a patu or mere
taua	war party
taupuhi	the bunches of feathers sometimes attached to tewhatewha and other weapons; also called puhipuhi
tauri kura	a band of red kaka feathers attached to the shaft of a taiaha
tipi	a straight jab from a weapon such as a patu paraoa or kotiate
toa	warrior
toki	adze
tuiri	a drill used on stone; also called horete
upoko	head of a taiaha
utu	satisfaction or revenge
waka taua	war canoe
whakawhiti	finely carved band on a pouwhenua
whare	house

SELECTED BIBLIOGRAPHY

Beaglehole, J.C. (ed.), *The Journals of Captain James Cook on his Voyages of Discovery,* vol. 1. Cambridge University Press, Cambridge, 1955.

Beattie, J.H. (A. Anderson, ed.), *Traditional Lifeways of the Southern Maori.* University of Otago Press, Dunedin, 1994.

Best, E., *The Maori.* Board of Maori Ethnological Research, Wellington, 1924.

——— (J. Evans, ed.), *Notes on the Art of War.* Reed, Auckland, 2001.

Buck, Sir P., *The Coming of the Maori.* Whitcoulls/Maori Purposes Board, Wellington, 1987 (2nd edn).

Hamilton, A., *Maori Art.* New Zealand Institute, Dunedin, 1897.

Maning, F.E., *Old New Zealand.* Viking, Auckland, 1987.

Phillipps, W.J., *Maori Life and Custom.* A.H. & A.W. Reed, Wellington, 1966.

Reedy, H.G., 'Te Tohu-a-Tuu (The Sign of Tuu): A Study of the Warrior Arts of the Maori'. PhD thesis, Massey University, Palmerston North, 1996.

Salmond, A., *Two Worlds: First meetings between Maori and Europeans, 1642–1772.* Viking, Auckland, 1991.

Shortland, E., *Southern Districts of New Zealand.* Capper Press, Christchurch, 1974.

Williams, H.A., *Dictionary of the Maori Language.* GP Books, Wellington, 1991 (7th edn).